HRJC

Distinctions in Nature

Herbivores and Carnivores Explained

Shirley Duke

New York

Published in 2017 by Cavendish Square Publishing, LLC
243 5th Avenue, Suite 136, New York, NY 10016

Copyright © 2017 by Cavendish Square Publishing, LLC

First Edition

No part of this publication may be reproduced, stored in a retrieval system, or transmitted in any form or by any means—electronic, mechanical, photocopying, recording, or otherwise—without the prior permission of the copyright owner. Request for permission should be addressed to Permissions, Cavendish Square Publishing, 243 5th Avenue, Suite 136, New York, NY 10016. Tel (877) 980-4450; fax (877) 980-4454.

Website: cavendishsq.com

This publication represents the opinions and views of the author based on his or her personal experience, knowledge, and research. The information in this book serves as a general guide only. The author and publisher have used their best efforts in preparing this book and disclaim liability rising directly or indirectly from the use and application of this book.

CPSIA Compliance Information: Batch #CS16CSQ

All websites were available and accurate when this book was sent to press.

Library of Congress Cataloging-in-Publication Data

Names: Duke, Shirley Smith, author.
Title: Herbivores and carnivores explained / Shirley Duke.
Description: New York : Cavendish Square Publishing, [2017] |
Series: Distinctions in nature | Includes bibliographical references and index. |
Description based on print version record and CIP data provided by
publisher; resource not viewed.
Identifiers: LCCN 2015051220 (print) | LCCN 2015049736 (ebook) |
ISBN 9781502617514 (ebook) | ISBN 9781502617750 (pbk.) |
ISBN 9781502617453 (library bound) | ISBN 9781502617590 (6 pack)
Subjects: LCSH: Food chains (Ecology)-Juvenile literature. |
Herbivores-Juvenile literature. | Carnivorous animals-Juvenile literature.
Classification: LCC QL756.5 (print) | LCC QL756.5 .D85 2017 (ebook)
DDC 577.16-dc23
LC record available at http://lccn.loc.gov/2015051220

Editorial Director: David McNamara
Editor: Kelly Spence
Copy Editor: Nathan Heidelberger
Art Director: Jeffrey Talbot
Designer: Stephanie Flecha
Production Assistant: Karol Szymczuk
Photo Research: J8 Media

The photographs in this book are used by permission and through the courtesy of: Jamen Percy/Shutterstock.com, cover (left); Cheryl Ann Quigley/Shutterstock.com, cover (right); leungchopan/Shutterstock.com, 4; Spencer Sutton/Science Source, 6; William Ervin/Science Source, 7; PeterVrabel/Shutterstock.com, 8; Aleksei Ruzhin/Shutterstock.com, 10 (left); Naturegraphica Stock/Shutterstock.com, 10 (right); File:Argiope_bruennichi_Cornacchiaia_1.jpg/Lucarelli/Wikimedia Commons, 11 (left); SCIEPRO/Science Photo Library/Getty Images, 11 (right); Ethan Tremblay/iStock/Thinkstock, 12; Roman Baiadin/iStock, 14; Sabena Jane Blackbird/Alamy Stock Photo, 15; Designua/Shutterstock.com, 16; JudiLen/iStock, 17; blickwinkel/Alamy Stock Photo, 18; Photoservice/iStock/Thinkstock, 19; enciktat/Shutterstock.com, 20 (left); Danita Delimont/Gallo Images/Getty Images, 20 (right); Michael Wick/Shutterstock.com, 22 (top); efendy/Shutterstock.com, 22 (bottom); Regien Paassen/Shutterstock.com, 24; KARNSTOCKS/Shutterstock.com, 26 (top); Palenque/Shutterstock.com, 26 (bottom); nico99/Shutterstock.com, 27.

Printed in the United States of America

Contents

Introduction: 5
Living Things and the Food Chain

Chapter 1: 9
Herbivores and Carnivores

Chapter 2: 13
Comparing Herbivores and Carnivores

Chapter 3: 21
Be a Food Chain Detective

Chapter 4: 25
Rule Breakers

Glossary 28

Find Out More 30

Index 31

About the Author 32

Giant pandas eat for about twelve hours each day. They mostly snack on the stalks, stems, and leaves of the bamboo plant.

Introduction: Living Things and the Food Chain

All living things need energy to survive. Some living things, like plants, make their own food. Animals, however, cannot make their own food. They get the energy they need to survive from the foods they eat. A **food chain** shows the kinds of foods different animals eat. It also explains how energy flows through an **ecosystem**.

Classifying Animals

Animals share many similarities and differences, including the types of foods they

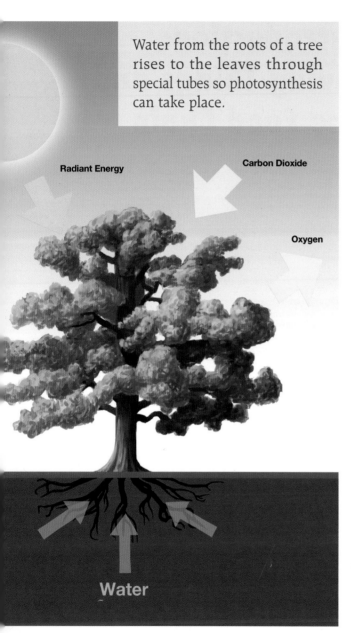

Water from the roots of a tree rises to the leaves through special tubes so photosynthesis can take place.

eat. Scientists use these similarities to **classify**, or group, animals. This branch of science is called **taxonomy**. Taxonomy helps scientists study and compare living things.

Plant Power

Plants are **producers**. They use water, air, and light from the sun to make their own food. The sunlight is soaked up by **chlorophyll**, which is the green coloring in plants. Plants turn this energy into food. This process is called **photosynthesis**.

Moving Up the Food Chain

All animals are **consumers**. They depend on producers for energy. But not all animals eat the same kinds of foods. Some get their energy by eating plants. These animals are classified as **herbivores**. Grasshoppers and elephants are herbivores.

Other animals, called **carnivores**, get energy by eating the animals that eat plants. They also hunt other meat-eating animals. Jackals and jaguars are carnivores.

Coyotes hunt mostly at night, but a hungry coyote will hunt for a snack during the day.

Living Things and the Food Chain

Young deer, called fawns, eat a varied diet of grass, leaves, weeds, bark, acorns, mushrooms, and moss as they grow up.

1 Herbivores and Carnivores

Some herbivores eat all the parts of a plant. Others are picky eaters. They might only eat the leaves, seeds, fruit, **nectar**, or flowers from a plant. Herbivores are **primary consumers** because they eat the plants that make the food.

Grazers and Browsers

Not all herbivores snack on the same foods. **Grazers** eat grass and small plants found near the ground. Other herbivores are **browsers**. Browsers eat leaves, plant shoots, and twigs. Elephants and moose are both browsers and grazers. Herbivores come in

Camels graze and browse on plants like grasses, twigs, and leaves. They avoid plants that are poisonous.

Lionesses hunt in groups and are able to capture larger prey like an antelope, wildebeest, or zebra.

all shapes and sizes. Larger herbivores need to eat more food than smaller ones.

A Meaty Meal

Animals that hunt and kill other animals are called **predators**. The animals they hunt are their **prey**. Most carnivores mainly hunt herbivores. Others will also eat other carnivores. Predators require lots of

Zoom In

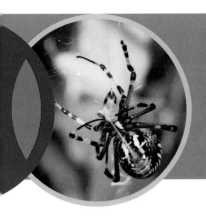

A spider is a carnivore. It spins a sticky web to capture its prey.

Blue whales feed on huge amounts of krill, a tiny shrimp-like animal.

food. They need plenty of energy to go out and hunt for their next meal.

Carnivores are **secondary consumers** because they eat herbivores, which are primary consumers. Most carnivores are small to medium in size. Large carnivores need to eat huge amounts of food. The largest carnivore in the world is the blue whale.

Fruit bats enjoy the sweet taste of fruits like watermelon, mango, and banana.

2 Comparing Herbivores and Carnivores

Herbivores like bison graze. Deer browse. However, some plants are low in **nutrients**, or the substances animals need to live. Herbivores must eat more plants to get enough nutrients. They spend most of the day eating. Herbivores like fruit bats eat nectar, flowers, and fruits. Giant pandas feed mainly on bamboo.

An Herbivore's Teeth

An herbivore's teeth are **adapted** for their diet. Their **molars**, teeth with a wide surface,

Zoom In

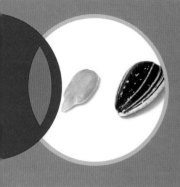

Small herbivores that eat seeds, like mice and chipmunks, survive on less food. Seeds are much richer in nutrients than other plant parts, such as the stem and leaves.

An elephant's tusks, which are part of its teeth, begin to grow when the animal is two or three years old. The tusks continue to grow for the rest of the elephant's life.

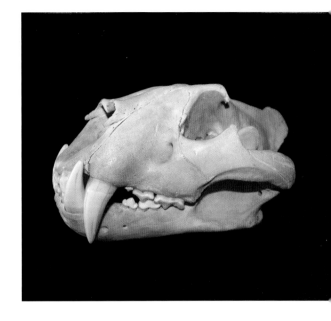

This tiger skull shows the sharp front teeth that carnivores use to hold prey. They use a scissor-like action to tear their food apart.

help them grind down plants. Ridges in these teeth also help herbivores chew grass and leaves. Some herbivores also have sharp front teeth called **incisors**. However, these animals do not use these sharp teeth to break down plants. Instead, they fight other animals with them or use them to strip bark or leaves off a tree or plant.

Biting Teeth

A carnivore's teeth are adapted to their diet, too. They have long, sharp teeth. These special teeth, called **canines**, are used to bite and tear into prey. Carnivores also have molars, but not as many as herbivores. Most carnivores eat their food in big chunks.

Built to Break Down

A cow is an herbivore. A cow has a special stomach chamber, called a **rumen**, that helps it break down grass. When a cow swallows its food after chewing, the grass travels to the rumen. There, bacteria break it down. The cow then brings the food back up to chew again. It is now called **cud**. The cow eats the cud and it travels through the animal's digestive system to break

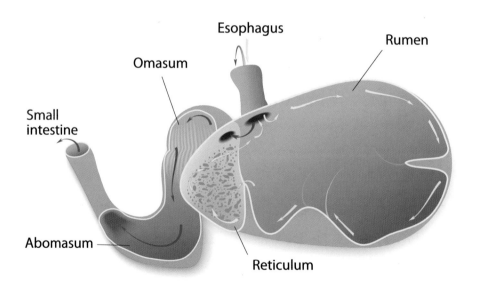

Food travels into the rumen, then back up the esophagus. After it is chewed and swallowed again, it travels on through the four-chambered stomach.

down further. Other animals, such as sheep, deer, and moose, also have a rumen.

Built to Hunt

Carnivores like wolves and hyenas have powerful jaws for hunting. They also have sharp claws. Birds that eat meat have a sharp, spiked beak and claws called

An osprey uses its sharp talons to grasp a fish, then carries it away to feed in a safe place.

Zoom In

The wolves in Yellowstone National Park were once thought to be harmful. Their main food source was elk. In 1926, humans hunted all the wolves until they were gone. This created an overpopulation of elk. The elk ate so many trees that they didn't grow back. This upset the ecosystem's balance. After wolves were introduced back to the park in 1995, the food chain slowly improved.

talons for tearing. They also have good eyesight for spotting prey.

Keeping the Balance

In an ecosystem, every living thing has a job to do. The food chain helps keep an ecosystem in balance.

In an unbalanced ecosystem, a herd of zebras must compete with other animals for food.

Carnivores help keep the herbivore population down. If the carnivores do not eat enough herbivores, there are too many herbivores. The herbivores eat the grass faster than it can grow. Before long, the herbivores run out of food. Then the carnivores lose their food source and die.

1. Bengal tigers are found mostly in India and other parts of Asia.

2. A capybara is the largest rodent in the world.

3 Be a Food Chain Detective

Be a food chain detective! Use the pictures (*left*) and clues (*below*) to classify the animals as herbivores or carnivores.

1. Think about the shape of this tiger's teeth. Do they look like teeth used for browsing and chewing tough bark or grass? Or do they look more like teeth made for gripping and tearing flesh? Is this animal an herbivore or carnivore?

2. Study this capybara's teeth. Notice its long incisors. Are these the teeth of an herbivore or carnivore?

Be a Food Chain Detective   21

3. The European golden eagle lives up to thirty-eight years in the wild.

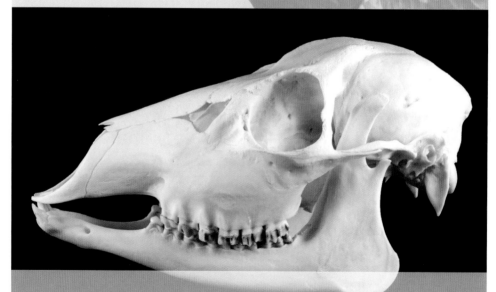
4. Scientists learn a lot about an animal's diet by studying its skull.

3. Look at the shape of this European golden eagle's beak. Do you think this bird eats plants or animals?

4. Examine the teeth in the jaw of this skull. Look at how they are shaped. Are they pointed or flat? What do you think this animal eats? Is it a carnivore or herbivore?

Answer Key:

1. Tigers are carnivores. They have very large canine teeth and sharp, pointed claws.

2. Capybaras are herbivores. They eat water plants and grasses and are found in Central and South America.

3. The European golden eagle is a carnivore. These birds need a large area to hunt for prey.

4. This is a deer skull. These animals are herbivores and have rumens to help break down their food.

A black bear's diet mainly consists of plant matter. This black bear is enjoying a feast of buffalo berries.

4 Rule Breakers

Not all animals are herbivores or carnivores. Animals called **omnivores** eat both plants and meat. They are an important part of the food chain. Omnivores are well suited for survival. They can change what they eat based on what is available. Omnivores like the black bear can live off plant matter like berries, roots, and fruits if they can't find meat.

Humans are omnivores. Our teeth are suited to eating plants and meat. Humans have sharp canines for chewing meat. We also have flat molars and incisors to break down fruits, vegetables, and grains. Some humans choose to not eat meat. They are called vegetarians.

Decomposers and Scavengers

Decomposers and **scavengers** break down plant and animal matter. They also break down waste. Scavengers, such as vultures, find dead matter and eat it. Decomposers, such as worms, finish what the scavengers

Humans have molars, incisors, and canines to break down both plant and animal matter.

When vultures feed on animal remains, they help keep the surrounding area clean.

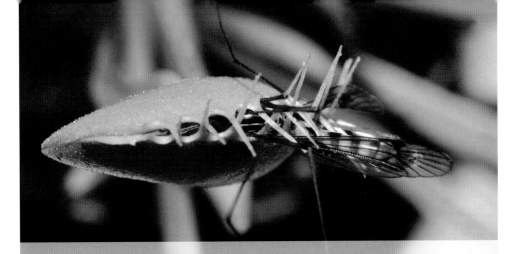

The Venus fly trap has sensitive hairs that signal to the plant to snap shut when food lands on it. Digestive juices dissolve the trapped insect.

do not want. Both decomposers and scavengers help nutrients return to the food chain. These nutrients help plants grow, creating new food for herbivores to eat.

Hungry Plants

Some plants are carnivorous. They have adapted to trap insects and even small animals. The trapped prey becomes the plant's next meal. The plant makes special chemicals to help break down the food. Some meat-eating plants trap food by closing their leaves around it. Others are covered in a sticky substance that grabs hold of their prey. Venus flytraps, sundews, and bladderworts are all carnivores.

Glossary

adapted To have changed over a long period of time.
browsers Herbivores that eat leaves, shoots, and twigs.
canines Pointed teeth used to grip and tear food.
carnivores Animals that eat meat.
chlorophyll The green coloring in plants that helps them make their own food.
classify To put things that are alike into groups.
consumers Things that eat or use up something.
cud Food that an animal brings back up from its rumen to chew again.
decomposers Life-forms that break down dead matter into nutrients that return to the soil.
ecosystem A community made up of living things.
food chain The ways animals depend on plants and other animals for getting food.
grazers Herbivores that eat plants near the ground.
herbivores Animals that eat plant matter.
incisors Front teeth used to cut and gnaw.

molars Flat teeth used to grind food, especially plant matter.

nectar A sugary liquid formed in flowers.

nutrients Substances that all living things need to grow.

omnivores Animals that eat both plants and meat.

photosynthesis The process by which plants convert light, air, and water into energy.

predators Animals that hunt and kill their food.

prey Animals that are hunted for food.

primary consumers Animals that eat plants.

producers Organisms that make their own food.

rumen A special stomach found in some herbivores that allows them to bring up cud and chew it again.

scavengers Animals that eat dead and decaying animals.

secondary consumers Animals that feed on primary consumers.

taxonomy The science of grouping living things by their similarities.

Find Out More

Books

Kalman, Bobbie. *What Is a Herbivore?* Big Science Ideas. New York: Crabtree Publishing, 2008.

Lowery, Lawrence F. *What Does an Animal Eat?* I Wonder Why. Arlington, VA: NSTA Kids, 2013.

Owen, Ruth. *How Do Meat-Eating Plants Catch Their Food?* The World of Plants. New York: PowerKids Press, 2015.

Websites

Nature Unleashed: You Eat What?!
mdc.mo.gov/sites/default/files/resources/2010/09/9959_6937.pdf
Learn more about a food chain in an ecosystem.

The Wildlife Web 2: Herbivores and Carnivores
video.nhptv.org/video/1491185781
Watch this video from PBS to learn all about the food chain.

Index

Page numbers in **boldface** are illustrations.

adapted, 13, 15, 27
browsers, 9, **10**, 13, 21
canines, 15, **15**, 25, **26**
carnivores, 7, **7**, 9–11, **10**, **11**, 15, **15**, 17–19, **17**, **20**, 21, **22**, 23, 25, 27, **27**
chlorophyll, 6
classify, 6–7, 21
consumers, 7, 10–11
cud, 16
decomposers, 26–27
ecosystem, 5, 18, **19**
food chain, 5, 18, 21, 25, 27
grazers, 9, **10**, 13
herbivores, 7, **8**, 9–10, **10**, **12**, 13–17, **14**, 19, **20**, 21, **22**, 23, 25, 27
incisors, 15, 21, 25, **26**
molars, 13, 15, 25, **26**
nectar, 9, 13
nutrients, 13–14, 27
omnivores, **24**, 25–26, **26**
photosynthesis, 6, **6**
predators, 10–11
prey, 10–11, **10**, 15, 18, 27
primary consumers, 9, 10–11
producers, 6–7
rumen, 16–17, **16**, 23
scavengers, 26–27, **26**
secondary consumers, 11
taxonomy, 6

About the Author

Shirley Duke has written more than fifty books about science and nature, but writing about life science is her specialty. Duke has seen the food chain in action many times, but the best remembered is her time in Alaska. She saw herbivores, carnivores, and omnivores eating to prepare for the winter—all from the safety of the bus.